BOOK 2

Polyiamonds Folding

Folding Polyiamonds into Deltaheda
with 12 Faces or Less

Dr. Keh-Ming Lu

Book 2 - Polyiamonds Folding: Folding Polyiamonds into Deltaheda with 12 Faces or Less
Copyright © 2022 by Dr. Keh-Ming Lu

Control Number ISBN
Paperback: 978-1-77419-119-4
eBook: 978-1-77419-120-0

All rights reserved. No part of this book may be reproduced or transmitted, downloaded, distributed, reverse engineered, or stored in or introduced into any information storage and retrieval system, in any form or by any means, including photocopying and recording, whether electronic or mechanical, now known or hereinafter invented without permission in writing from the publisher.

Disclaimer: This is a work of nonfiction. No names have been changed, no characters invented, no events fabricated.

To order additional copies of this book, please contact:

MAPLE LEAF PUBLISHING INC.
www.mapleleafpublishinginc.com
3rd Floor 4915 54 St Red Deer,
Alberta T4N 2G7 Canada
General Inquiries & Customer Service:
Phone: 1-(403)-356-0255
Toll Free: 1-(888)-498-9380
Email: info@mapleleafpublishinginc.com

Introduction

Counting 2n-Deltahedra (1/3)

This book focused on folding polyiamonds into deltahedra with 12 faces or less. A deltahedron is a polyhedron whose faces are all equilateral triangles. There is only one deltahedron with four faces: the tetrahedron. Likewise, there is only one deltahedron with six faces: the triangular bipyramid. There are two with eight faces: the octahedron and the biaugmented tetrahedron (not convex). And five with 10 faces: the pentagonal bipyramid, the augmented octahedron (contains coplanar faces), the three triaugmented tetrahedra. And 13 with 12 faces with a help from OEIS (The On-Line Encyclopedia of Integer Sequences) provides the counts of polyiamonds.

Counting 2n-Deltahedra (2/3)

I applied OEIS A000577 to proof the count of 2n faces deltahedra. I used 3 4-iamonds (A000577(4)=3) to proof only one deltahedron with four face. I used 12 6-iamonds (A000577(6)=12) proof only one deltahedron with six faces. Also, I used 66 8-iamonds (A000577(8)=66) proof only two deltahedron with eight faces. In the same manners, I used 448 10-iamonds (A000577(10)=448) proof only five (5) deltahedron with ten faces. I have spent two years (year 2018 -2020) and used 3334 12-iamonds (A000577(12)=3334) to construct and proof thirteen (13) deltahedral graphs with twelve faces.

Counting 2n-Deltahedra (3/3)

They are D12(0,4,4,0), D12(1,3,3,1), D12(1,4,1,2), D12(2,0,6,0), D12(2,1,4,1), D12(2,2,2,2)1, D12(2,2,2,2)2, D12(2,2,3,0,1), D12(2,3,1,1,1)1, D12(2,4,0,0,2), D12(3,1,1,3), D12(3,1,2,1,1), D12(4,0,0,4). Where D12 stands for a deltahedron with 12 faces. For example, D12(0,4,4,0) denote that 12-deltahedron with 8 vertices: 4 of degree 4 and 4 of degrees 5 and none of degree 3 and 6.

Polyiamond (1/2)

A polyiamond is a polyform whose base form is an equilateral triangle. In geometry, an equilateral triangle is a triangle in which all three sides are equal. An equilateral triangle is also equiangular; that is, all three internal angles are each 60°.

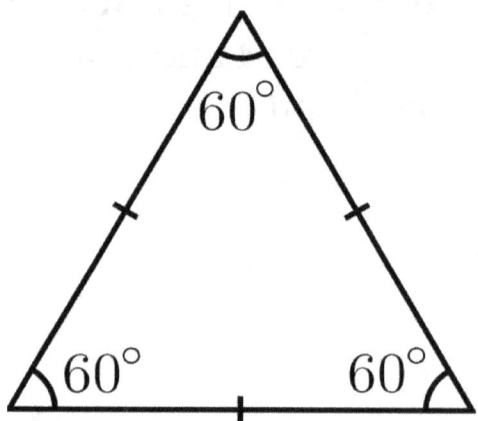

Polyiamond (2/2)

The number of n-iamonds for n = 1, 2, 3, ... is: 1, 1, 1, 3, 4, 12, 24, 66, 160, 448, 1186, 3334, ... (sequence A000577 in the OEIS). Sequence A000577 is number of polyamonds with n cells (turning over is allowed, holes are allowed, must be connected along edges). We only use even numbers of Polyiamond: Tetriamond (3), Hexiamond (12), Octiamond (66), Deciamond (448), Dodeciamond (3334)

Deltahedron (1/3)

A deltahedron is a polyhedron whose faces are all equilateral triangles. The name is taken from the Greek majuscule delta (Δ), which has the shape of an equilateral triangle. There are infinitely many deltahedra, but of these only eight (8) are convex, having 4, 6, 8, 10, 12, 14, 16 and 20 faces. Three are regular polyhedra: tetrahedron (4 faces), octahedron (8), icosahedron(20). Five are Johnson solids: triangular bipyramid (6), pentagonal bipyramid (10), snub disphenoid (12), triaugmented triangular prism (14), gyroelongated square bipyramid (16)

Deltahedron (2/3)

All deltahedra with face 12 or less can be folded from even numbers of Polyiamonds: Tetriamonds (3), Hexiamonds (12), Octiamond (66), Deciamond (448), Dodeciamond (3334). Folding Tetriamonds (3), Hexiamonds (12), and Octiamonds (66) into deltahedra with face 4, 6 and 8 can be done in weeks. However, folding Deciamond (448) and Dodeciamond (3334) into 10-deltahedra, or Decahedra, and 12-deltahedra can easily spent more than two years to finish.

Deltahedron (3/3)

The purpose of this book is to save your time although you may insist following my footsteps to obtain the same result. The Book consists of an Introduction and six (6) Chapters: Chapter 1 Tetrahedron; Chapter 2 Hexahedron; Chapter 3 Octahedra; Chapter 4 Decahedra; Chapter 5 12-Deltahedra; Chapter 6 Appendices.

Prof. Keh-Ming Lu

Chapter 1

Tetrahedron

Tetrahedron (1/6)

Number of Tetrahedron is 1, denoted D4(4000). Tetrahedron has 4 faces (F), 6 edges (E), and 4 vertices (V). Tetrahedron D4(4000) satisfy Euler's Formula: $F+V=E+2$; $F+V=4+4=8$, $E+2=6+2=8$. D stands for Deltahedron, 4 stands for 4 faces. The numbers inside the parenthesis are the Degree of the deltahedron, in the sequence of Degree of 3, 4, 5, 6. It means 4 vertices degree of 3. Degree is the number of edges of the vertex.

Tetrahedron (2/6)

D4(4000) is has 4 vertices of Degree of 3. The number of Tetriamonds is: 3, ie a(4) of sequence A000577 in the OEIS). Construct a Tetrahedron D4(4000) from one of 3 Tetriamonds.

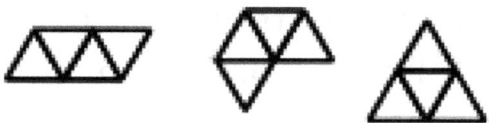

tetriamonds

Tetrahedron (3/6)

We name them in sequence, Tetriamond (1), Tetriamond (2) and Tetriamond (3).

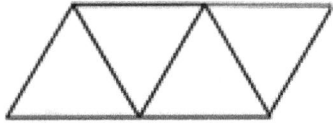

Tetriamond (1)

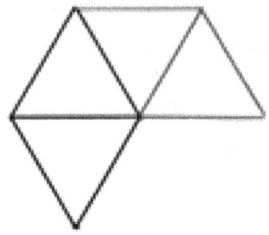

Tetriamond (2)

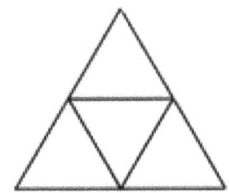

Tetriamond (3)

Tetrahedron (4/6)

Among 3 Tetriamonds, Tetriamond (2) that can not be folded into D4(4000). This is due to two or more faces of Tetriamond (2) glued together in process of construction.

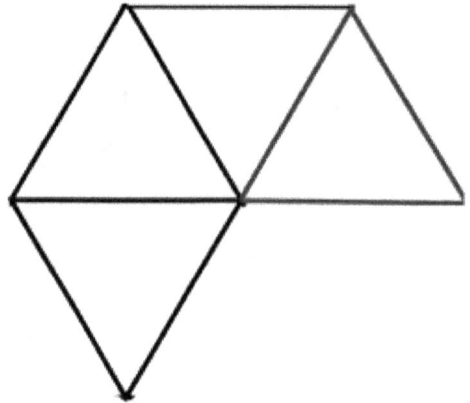

Tetriamond(2)

Tetrahedron (5/6)

Construct D4(4000) from Tetriamond (1). The Tetriamond (1) has three pairs of edges. Gluing edge 1 and edge 4 denote as (e1 e4) Similarly gluing edge 2 and edge 3, edge 5 and edge 6, denoted as (e2 e3) and (e5 e6) respectfully.

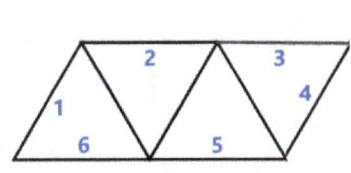

Numbered
Tetriamond(1)

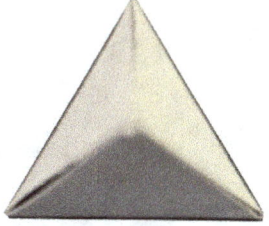

Tetrahedron

D4(4000)=(e1 e4)(e2 e3)(e5 e6)

Tetrahedron (6/6)

Construct D4(4000) from Tetriamond (3).

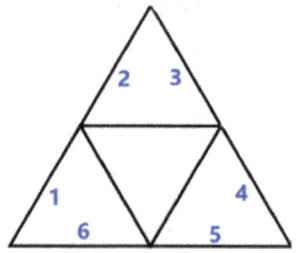

Numbered Tetriamond(3)

Tetrahedron

D4(4000)=(e1 e2)(e3 e4)(e5 e6)

CHAPTER 2

Hexahedron

Hexahedron (1/12)

Number of Hexahedra is 1, denoted D6(2300). It also names Triangular Dipyramid (J12). Hexahedron has 6 faces (F), 9 edges (E), and 5 vertices (V). Hexahedron satisfy Euler's Formula: F+V=E+2; F+V=6+5=11, E+2=9+2=11; D stands for Deltahedron, 6 stands for 6 faces. The numbers inside the parenthesis are the Degree of the deltahedron, in the sequence of Degree of 3, 4, 5, 6. It means 2 vertices degree of 3 and 3 vertices degree of 4. Degree is the number of edges of the vertex.

Hexahedron (2/12)

D6(2300) is an Hexahedron and has 2 vertices of Degree of 3 and 3 vertices of Degree of 4. Hexahedron has 5 vertices. Construct a hexahedron D6(2300) from one of 12 Hexiamonds.

Hexahedron (3/12)

The number of Hexiamonds is: 12, ie a(6) of sequence A000577 in the OEIS). The 12 hexiamonds are illustrated above. They are given the names: bar, crook, crown, sphinx, snake, yacht, chevron, signpost, lobster, hook, hexagon, and butterfly.

We name them in sequence, Hexiamond (1), ... Hexiamond (12).

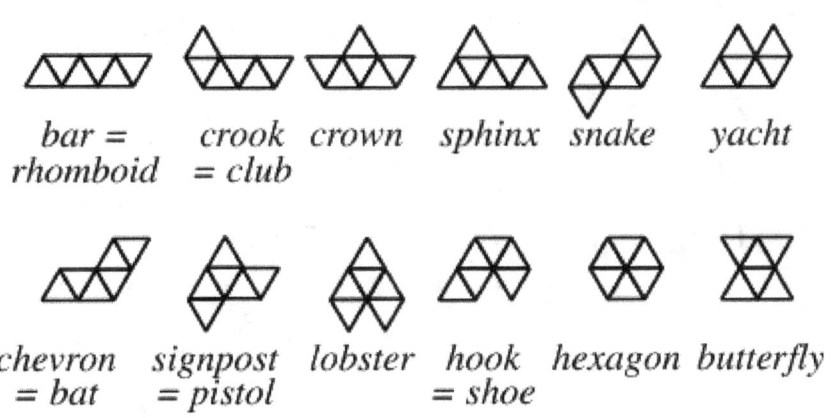

bar = rhomboid   crook = club   crown   sphinx   snake   yacht

chevron = bat   signpost = pistol   lobster   hook = shoe   hexagon   butterfly

Hexahedron (4/12)

Among 12 Hexiamonds, there are Hexiamond (6), Hexiamond (9), Hexiamond (10) and Hexiamond (11) that can not construct D6(2300). This is due to two or more faces of Hexiamonds glued together in process of construction.

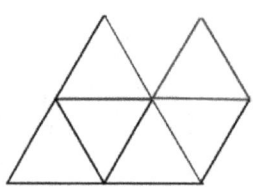

Hexiamond (6)

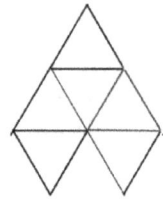

Hexiamond (9)

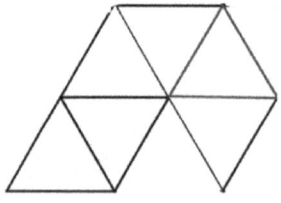

Hexiamond (10)

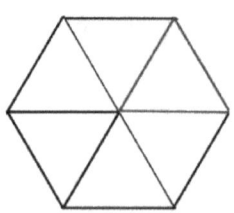

Hexiamond (11)

Hexahedron (5/12)

Construct D6(2300) from Hexiamond (1).

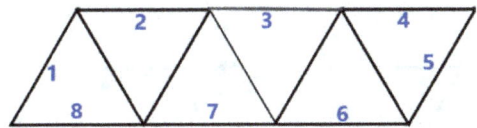

Hexiamond (1)

D6(2300) =(e1 e6)(e2 e5)(e3 e4)(e7 e8)

Hexahedron (6/12)

Construct D6(2300) from Hexiamond (2).

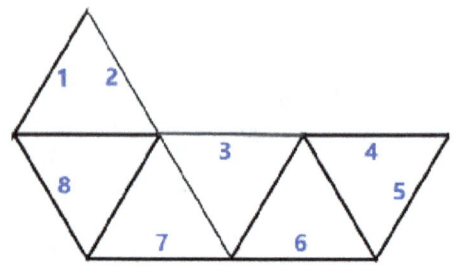

Hexiamond (2)

D6(2300) =(e1 e4)(e2 e3)(e5 e8)(e6 e7)

Hexahedron (7/12)

Construct D6(2300) from Hexiamond (3).

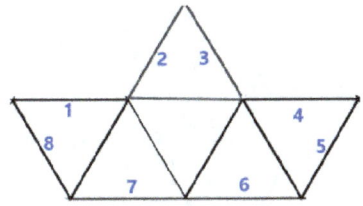

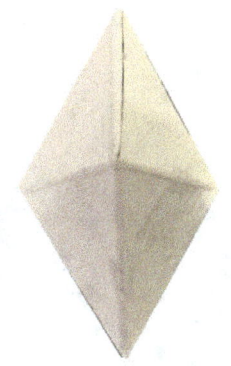

Hexiamond (3)

D6(2300) =(e1 e2)(e3 e4)(e5 e8)(e6 e7)

Hexahedron (8/12)

Construct D6(2300) from Hexiamond (4).

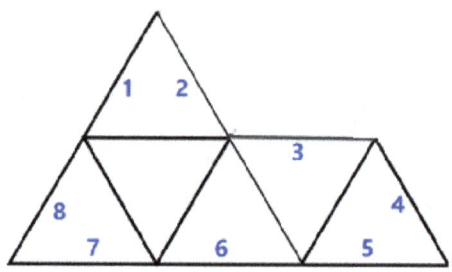

Hexiamond (4)
D6(2300) =(e1 e4)(e2 e3)(e5 e8)(e6 e7)

Hexahedron (9/12)

Construct D6(2300) from Hexiamond (5).

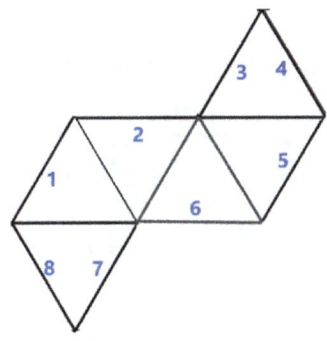

Hexiamond (5)

D6(2300) =(e1 e4)(e2 e3)(e5 e8)(e6 e7)

Hexahedron (10/12)

Construct D6(2300) from Hexiamond (7).

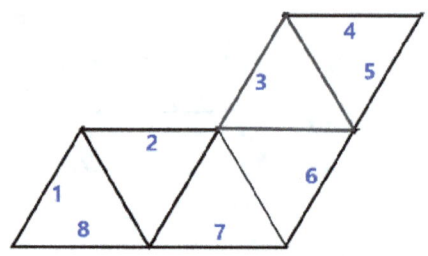

Hexiamond (7)

D6(2300) =(e1 e4)(e2 e3)(e5 e6)(e7 e8)

Hexahedron (11/12)

Construct D6(2300) from Hexiamond (8).

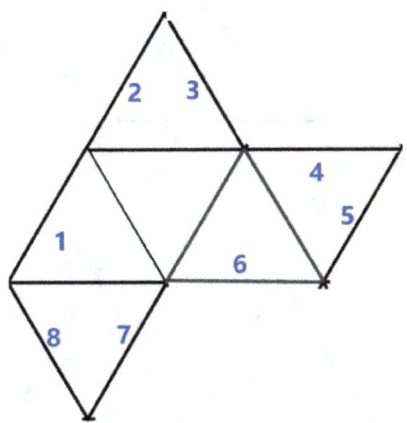

Hexiamond (8)

D6(2300) =(e1 e2)(e3 e4)(e5 e8)(e6 e7)

Hexahedron (12/12)

Construct D6(2300) from Hexiamond (12).

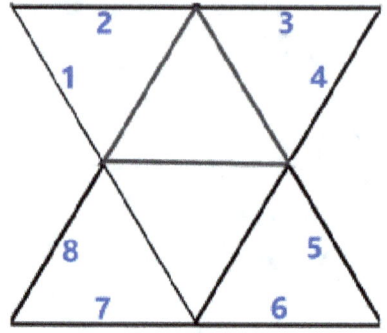

Hexiamond (12)

D6(2300) =(e1 e8)(e2 e3)(e4 e5)(e6 e7)

Octahedra

Octahedron (1/12)

Number of Octahedra is 2, denoted D8(0600) and D8(2220). Octahedron has 8 faces (F), 12 edges (E), and 6 vertices (V). All Octahedron satisfy Euler's Formula: $F+V=E+2$; $F+V=8+6=14$, $E+2=12+2=14$; D stands for Deltahedron, 8 stands for 8 faces. The numbers inside the parenthesis are the Degree of the deltahedron, in the sequence of Degree of 3, 4, 5, 6. Degree is the number of edges of the vertex.

Octahedron (2/12)

D8(0600) is an octahedron and has 6 vertices of Degree of 4. D8(2220) is also an octahedron and has 3 sets of 2 vertices of Degree of 3, 4, 5. Octahedron has 6 vertices.

Cut all 3-dimension Octahedra into numbers of 2-dimension Octiamonds.

Octahedron (3/12)

The number of Octiamonds is: 66, ie a(8) of sequence A000577 in the OEIS. Refer to Appendice A3 for Octiamond (66).

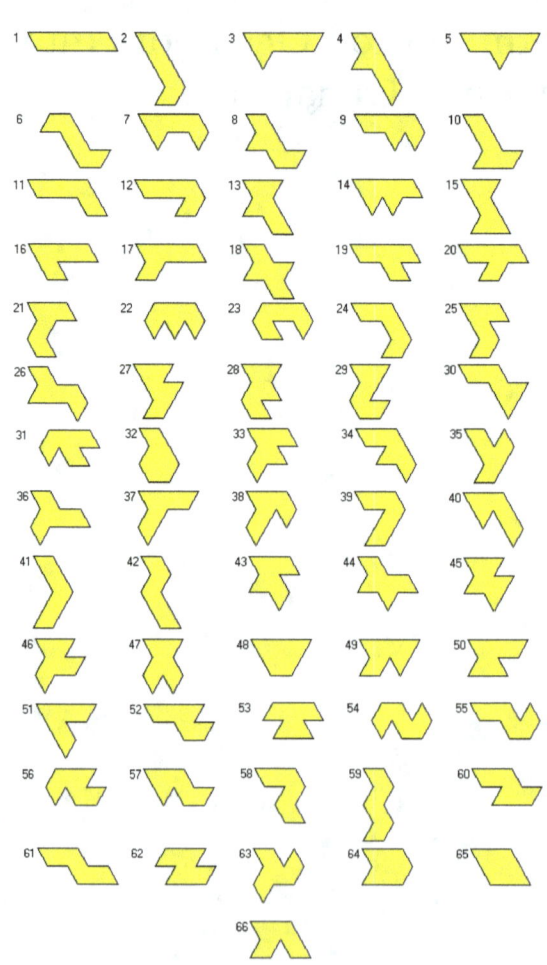

Octahedron (4/12)

Both Octiamond (32) and Octiamond (64) can not construct neither D8(2220) or D8(0600). This is due to two or more faces of Octiamonds glued together in process of construction.

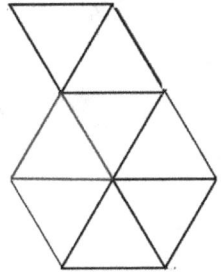

Octiamond(32)

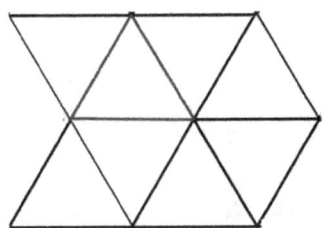

Octiamond(64)

Octahedron(5/12)

Having folded from 1 through 66 Octiamonds, we found only Octiamond (18), Octiamond (36), and Octiamond (61) each can construct both D8(2220) and D8(0600). Other 61 Octiamonds can only construct either D8(2220) and D8(0600).

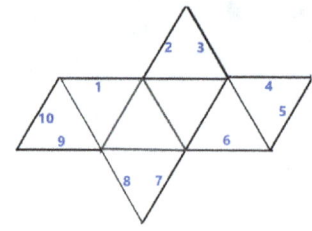

Octiamond(18)

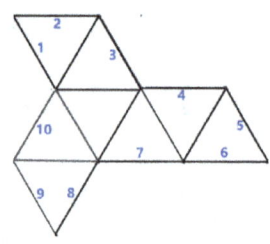

Octiamond(36)

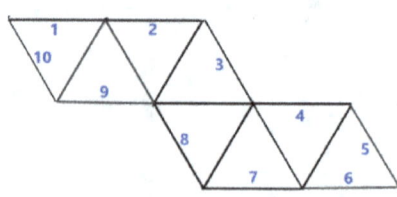

Octiamond(61)

Octahedron (6/12)

Only Octiamond (18) , Octiamond (36), and Octiamond (61) each can construct both D8(2220) and D8(0600).

D8(2220)

D8(0600)

Octahedron (7/12)

Construct D8(2220) from Octiamond(18).

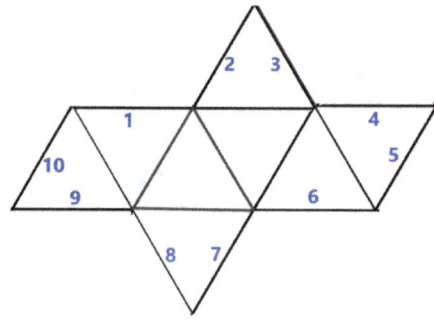

D8(2220)=(e1 e2)(e3 e10)(e4 e9)(e5 e8)(e6 e7).

Octahedron (8/12)

Construct D8(0600) from Octiamond(18).

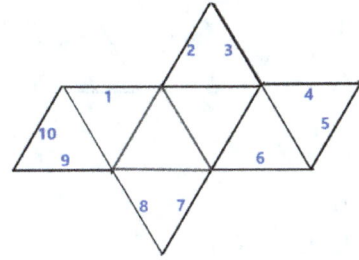

Octiamond (18)

D8(0600)

D8(0600)=(e1 e2)(e3 e4)(e5 e10)(e6 e7)(e8 e9)

Octahedron (9/12)

Construct D8(2220) from Octiamond(36).

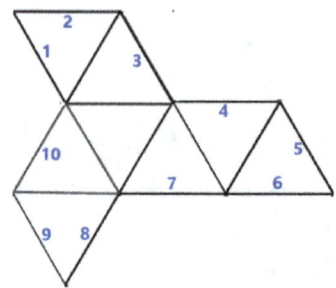

Octiamond (36)

D8(2220)

D8(2220)=(e1 e10)(e2 e9)(e3 e4)(e5 e8)(e6 e7)

Octahedron (10/12)

Construct D8(0600) from Octiamond(36).

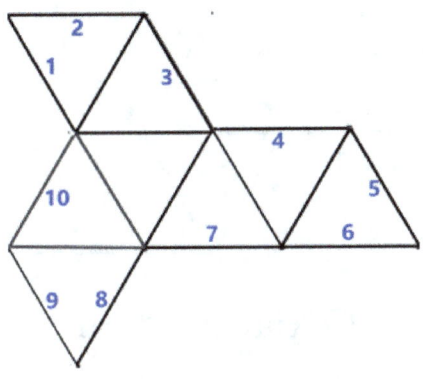

Octiamond (36)

D8(0600)

D8(0600)=(e1 e10)(e2 e5)(e3 e4)(e6 e9)(e7 e8)

Octahedron (11/12)

Construct D8(2220) from Octiamond(61).

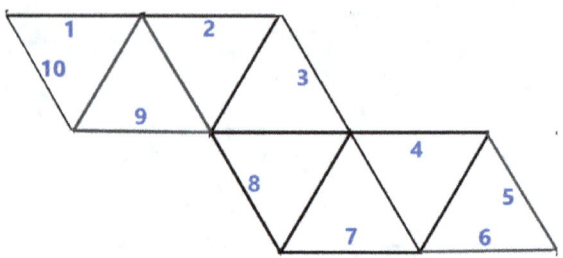

Octiamond (61)

D8(2220)

D8(2220)=(e1 e2)(e3 e4)(e5 e10)(e6 e7)(e8 e9).

Octahedron (12/12)

Construct D8(0600) from Octiamond(61).

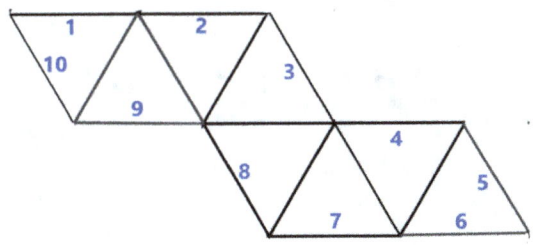

Octiamond (61)

D8(0600)

D8(0600)=(e1 e6)(e2 e5)(e3 e4)(e7 e10)(e8 e9)

Chapter 4

Decahedra

Deciamond (1/2)

Deciamond is a polyiamond made up of 10 equilateral triangles. The number of n-iamonds for n = 1, 2, 3, ... is: 1, 1, 1, 3, 4, 12, 24, 66, 160, 448, 1186, 3334, ... (sequence A000577 in the OEIS.. The number of a(10) of Sequence A000577 is 448. Refer to Appendice A4 for Deciamond (448). The name a(10) is named Deciamond.

Deciamond (2/2)

The number of Deciamond is 448.

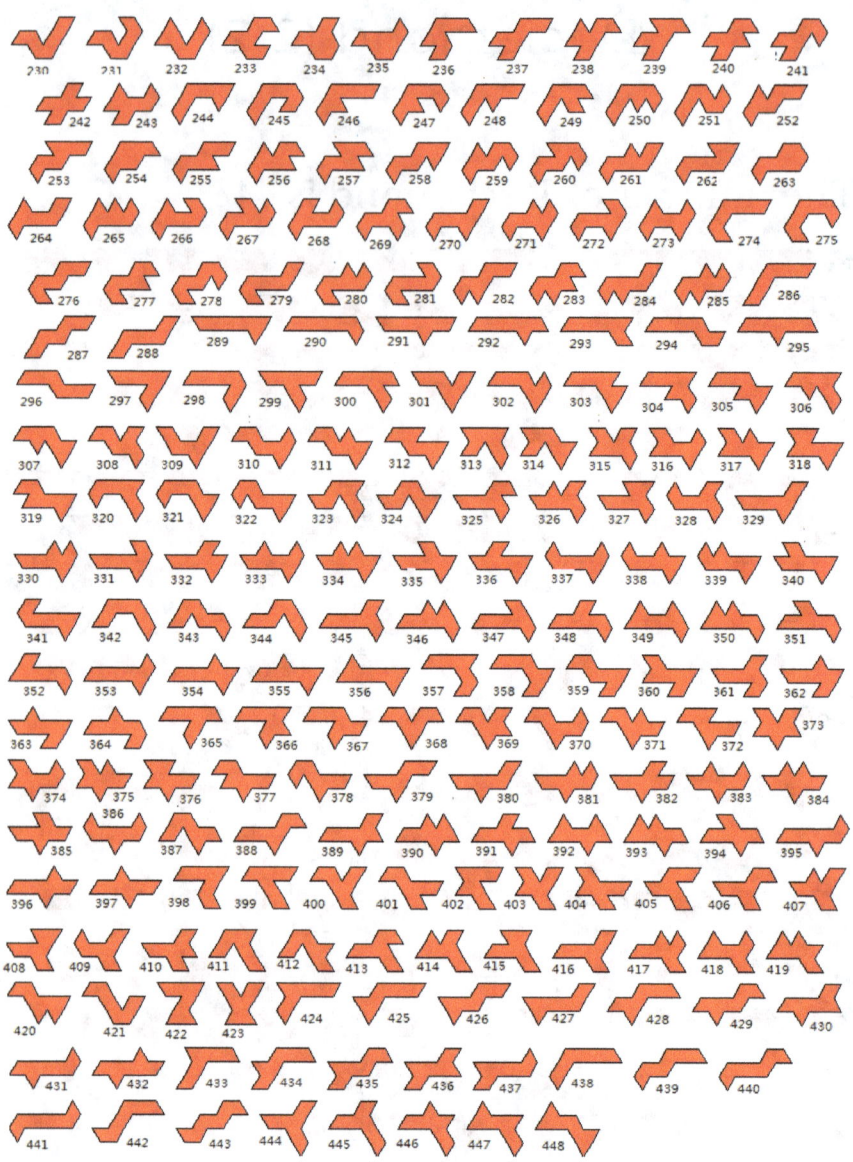

Decahedron (1/14)

A decahedron is a deltahedron with ten equilateral triangles faces. The number of Decahedron is 5. Construct a Decahedron from one of 448 Deciamonds. Having started Deciamond (1) and ended Deciamond (448), we have found 4 Decahedra: D10(0520), D10(2221), D10(2302), D10(3031) and 1 Non-deltahedral Graph : D10(1330). Also, all Deciamonds can be glued into at least one Decahedron.

Decahedron (2/14)

Having glued Deciamond (1) through Deciamond (448), there exists only Deciamond (318) and Deciamond (430) each can construct all 5 Decahedra.

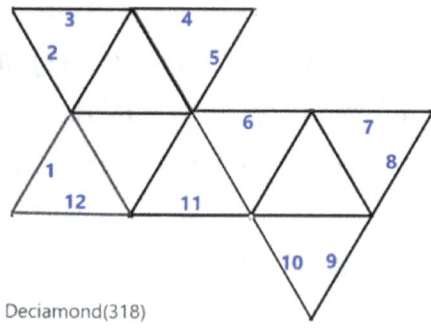

Deciamond (318)

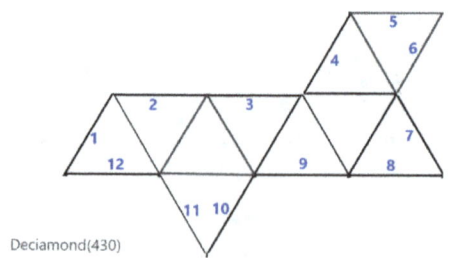

Deciamond (430)

Decahedron (3/14)

Cut 5 pieces of Deciamond(318) and Deciamond(430). Five (5) Decahedra are D10(0520), D10(1330), D10(2221), D10(2302), D10(3031).

Decahedron (4/14)

Construct D10(0520) from Deciamond (318).

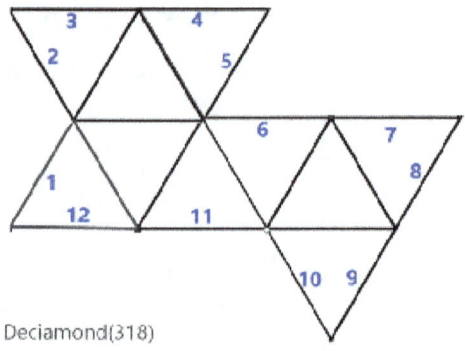

Deciamond (318)

D10(0520)=(e1 e2)(e3 e8)(e4 e7)(e5 e6)(e9 e12)(e10 e11)

Decahedron (5/14)

Construct D10(1330) from Deciamond (318).

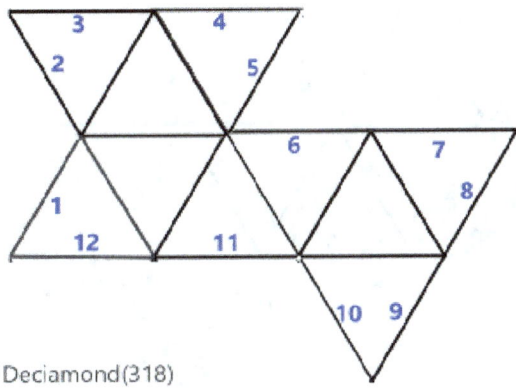

Deciamond(318)

D10(1330)= (e1 e8)(e2 e7)(e3 e4)(e5 e6)(e9 e12)(e10 e11)

Decahedron (6/14)

Construct D10(2221) from Deciamond (318).

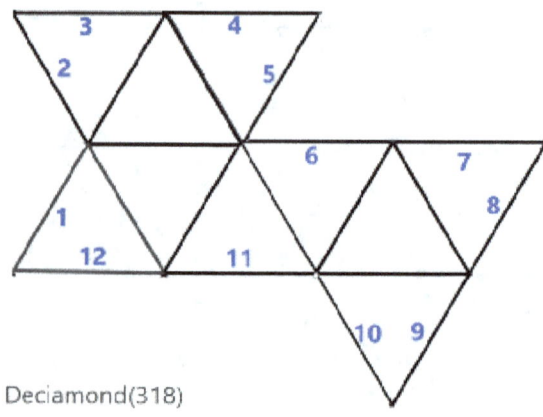

Deciamond (318)

D10(2221)= (e1 e2)(e3 e10)(e4 e7)(e5 e6)(e8 e9)(e11 e12)

Decahedron (7/14)

Construct D10(2302) from Deciamond (318).

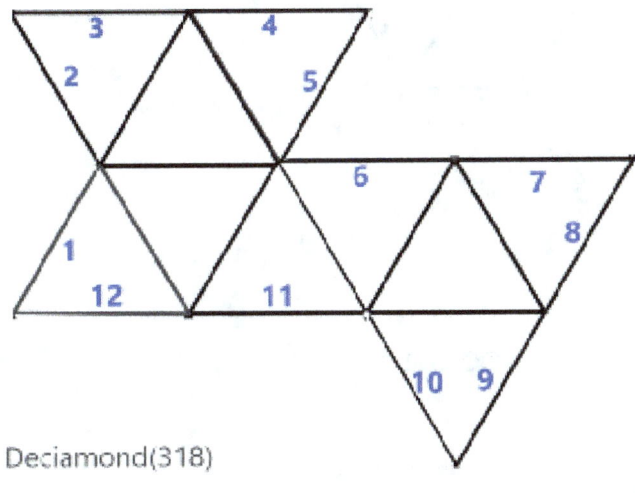

Deciamond (318)

D10(2302)= (e1 e2)(e3 e10)(e4 e9)(e5 e8)(e6 e7)(e11 e12)

Decahedron (8/14)

Construct D10(3031) from Deciamond (318).

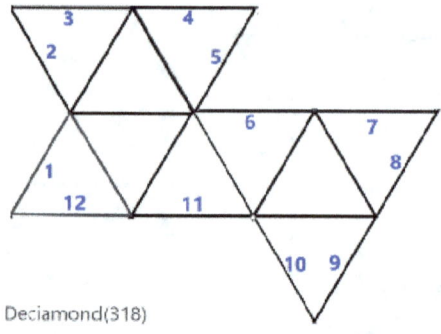

Deciamond (318)

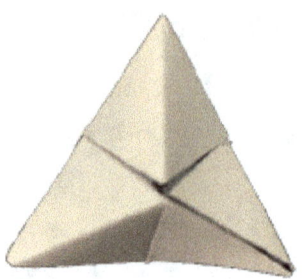

D10 (3031)

D10(3031)= (e1 e10)(e2 e7)(e3 e4)(e5 e6) (e8 e9)(e11 e12)

Decahedron (9/14)

Construct D10(0520) from Deciamond (430).

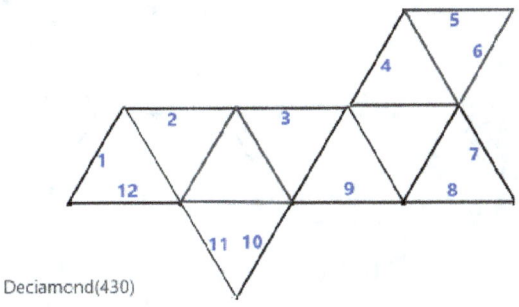

Deciamond(430)

D10(0520)= (e1 e8)(e2 e5)(e3 e4)(e6 e7)(e9 e10)(e11 e12)

Decahedron (10/14)

Construct D10(1330) from Deciamond (430).

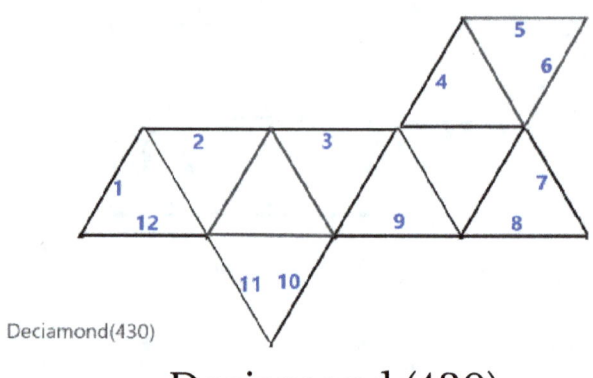

Deciamond (430)

D10(1330)= (e1 e6)(e2 e5)(e3 e4)(e7 e12)(e8 e11)(e9 e10)

Decahedron (11/14)

Construct D10(2221) from Deciamond (430).

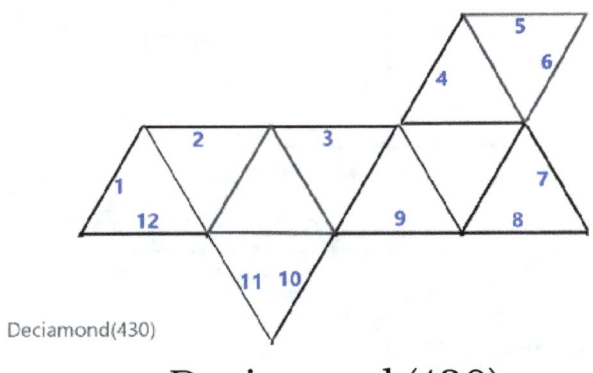

Deciamond (430)

D10(2221)= (e1 e6)(e2 e5)(e3 e4)(e7 e10)(e8 e9)(e11 e12)

Decahedron (12/14)

Construct D10(2302) from Deciamond (430).

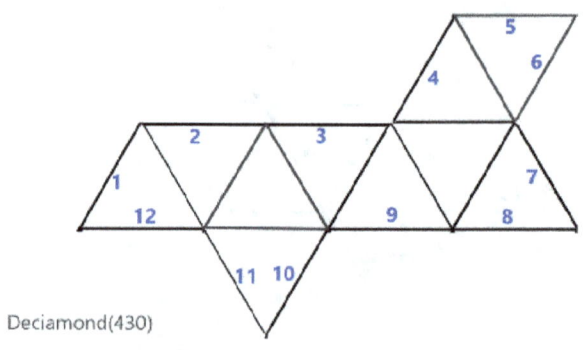

Deciamond (430)

D10(2302)= (e1 e4)(e2 e3)(e5 e10)(e6 e7)(e8 e9)(e11 e12)

Decahedron (13/14)

Construct D10(3031) from Deciamond (430).

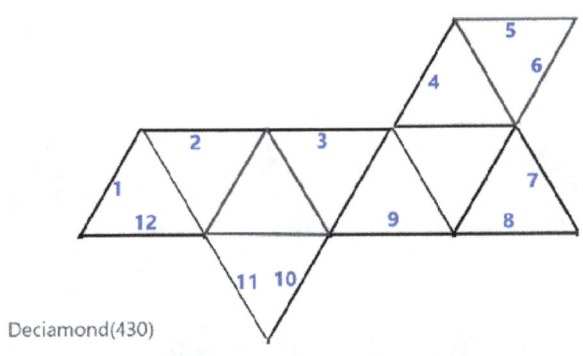

Deciamond (430)

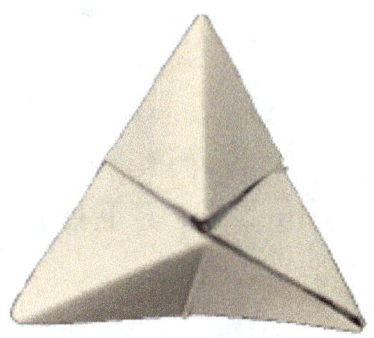

D10(3031)= (e1 e4)(e2 e3)(e5 e12)(e6 e11)(e7 e10)(e8 e9)

Decahedron (14/14)

Five (5) Decahedra graphs are D10(0520), D10(1330), D10(2221), D10(2302), D10(3031).

Decahedron has 10 faces (F), 15 edges (E), and 7 vertices (V). All five (5) Decahedra satisfy Euler's Formula: F+V=E+2, F+V=10+7=17, E+2=15+2=17

Chapter 5

12-Detahedron

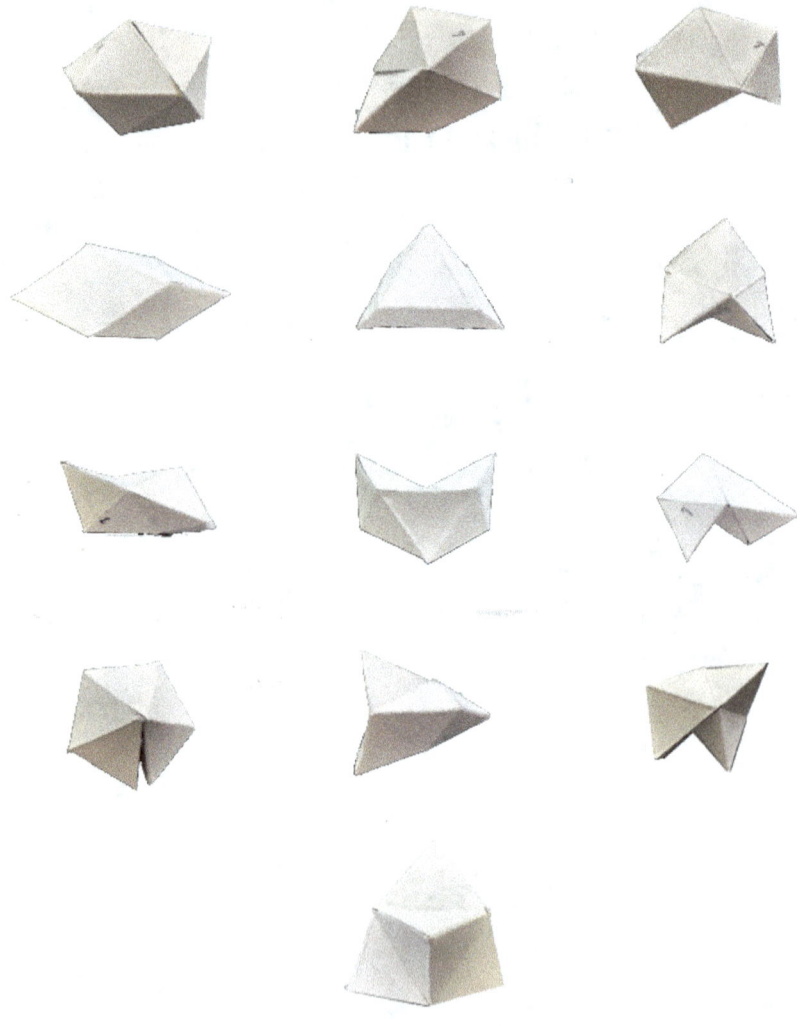

Dodecahedron (1/2)

In geometry, a dodecahedron or duodecahedron is any polyhedron with twelve flat faces. This Chapter is focused on Folding dodeciamond, or twelve equilateral triangles into 12 Faces of Deltahedra or 12-Deltahedron. The 12-Deltahedron are just one type of dodecahedron.

Dodeciamond (2/2)

Dodeciamond is a polyiamond made up of 12 equilateral triangles. The number of n-iamonds for n = 1, 2, 3, ... is: 1, 1, 1, 3, 4, 12, 24, 66, 160, 448, 1186, 3334, ... (sequence A000577 in the OEIS.. The number of a(12) of Sequence A000577 is 3334. Refer to Appendice A5 for 3334 Dodeciamonds. The name a (12) is named Dodeciamond.

12-Deltahedron (1/19)

In geometry, a 12-Deltahedron is a deltahedron with 12 equilateral triangles faces. The number of 12-Deltahedron is 13. Construct a 12-Deltahedron from one of 3334 Dodeciamonds. You may start folding Dodeciamond (1) and end Dodeciamond (3334). Having glued all dodeciamonds, I finished eight 400 pages Note Books in two years. We have found nine (9) 12-Deltahedron and 4 Non-deltahedral Graphs.

12-Deltahedron (2/19)

Both Dodeciamond (0432) and Dodeciamond (3171) can not construct none of 13 12-Deltahedra. This is due to two or more faces of Dodeciamond glued together in process of construction.

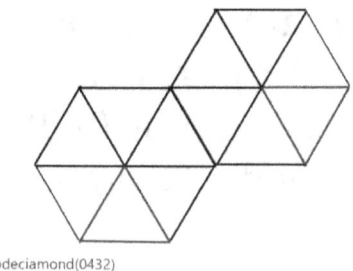
Dodeciamond(0432)

Dodeciamond (0432)

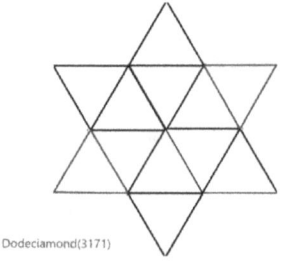
Dodeciamond(3171)

Dodeciamond (3171)

12-Deltahedron (3/19)

Combining Crease Patterns: Dodeciamond (2201) and Dodeciamond (2496) together can construct all 13 12-Deltahedra.

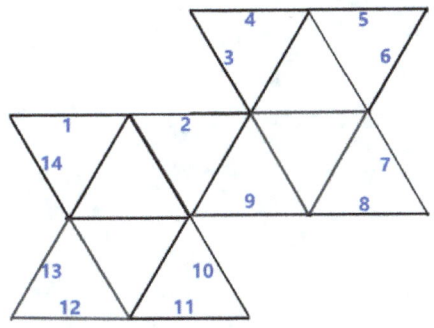

Dodeciamond (2201)

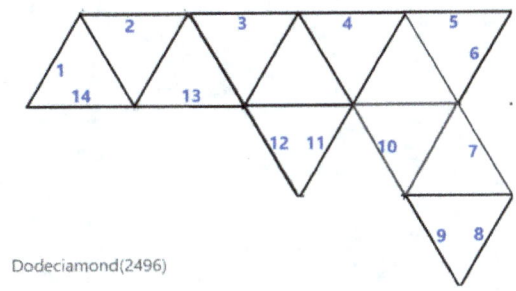

Dodeciamond (2496)

12-Deltahedron (4/19)

Cut 3 pieces of Dodeciamond(2201). Construct three (3) 12-Deltahedra: D12(4004), D12(0440), and D12(2060).

12-Deltahedron (5/19)

Construct D12(4004) from Dodeciamond (2201)

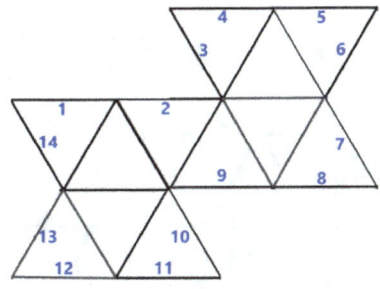

Dodeciamond(2201)

D12(4004)= (e1 e2)(e3 e14)(e4 e5)(e6 e13)(e7 e10) (e8 e9) (e11 e12)

12-Deltahedron (6/19)

Construct D12(0440) from Dodeciamond (2201).

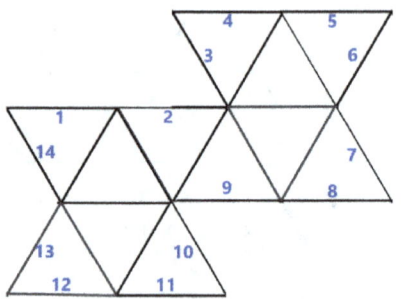

Dodeciamond(2201)

D12(0440)= (e1 e4)(e2 e3)(e5 e14)(e6 e13)(e7 e12) (e8 e11) (e9 e10)

12-Deltahedron (7/19)

Construct D12(2060) from Dodeciamond (2201).

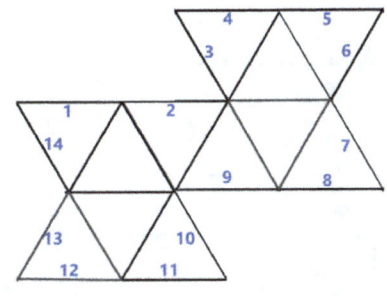

Dodeciamond(2201)

Dodeciamond (2201)

D12(2060)= (e1 e6)(e2 e3)(e4 e5)(e7 e14)(e8 e13) (e9 e10) (e11 e12)

12-Deltahedron (8/19)

Construct D12(1331) from Dodeciamond (2496).

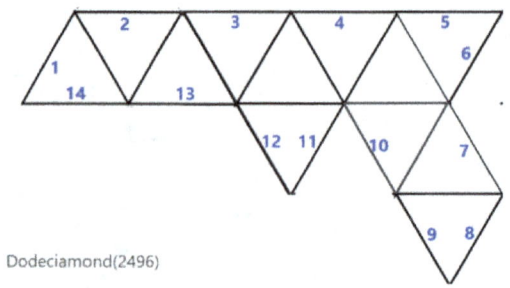

Dodeciamond (2496)

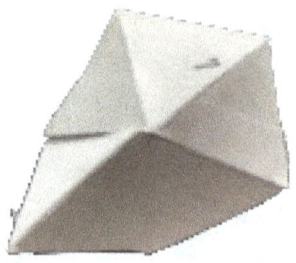

- D12(1331)= (e1 e4)(e2 e3)(e5 e14)(e6 e7)(e8 e13) (e9 e12) (e10 e11)

12-Deltahedron (9/19)

Construct D12(23111) from Dodeciamond (2496).

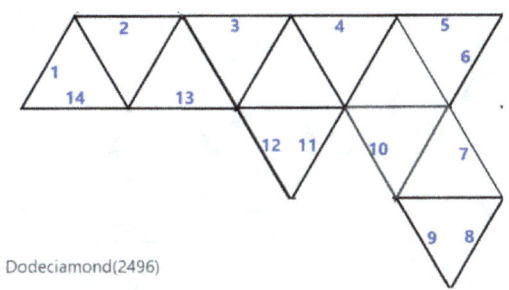

Dodeciamond (2496)

D12(23111)= (e1 e4)(e2 e3)(e5 e14)(e6 e7)(e8 e11) (e9 e10) (e12 e13)

12-Deltahedron (10/19)

Construct D12(22301) from Dodeciamond (2496).

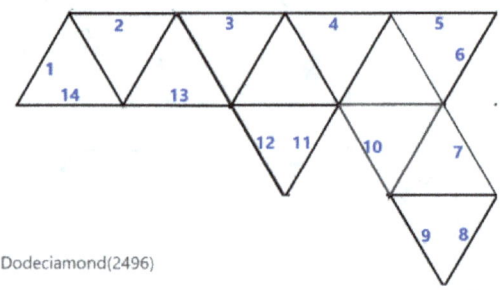

Dodeciamond (2496)

D12(22301)= (e1 e4)(e2 e3)(e5 e8)(e6 e7)(e9 e14) (e10 e11)(e12 e13)

12-Deltahedron (11/19)

Construct D12(2141) from Dodeciamond (2496).

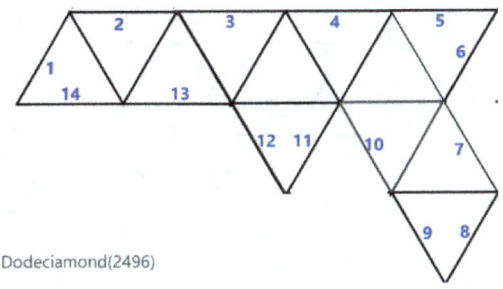

Dodeciamond(2496)

Dodeciamond (2496)

D12(2141)= (e1 e6)(e2 e3)(e4 e5)(e7 e14)(e8 e13) (e9 e12) (e10 e11)

12-Deltahedron (12/19)

Construct D12(3113) from Dodeciamond (2496).

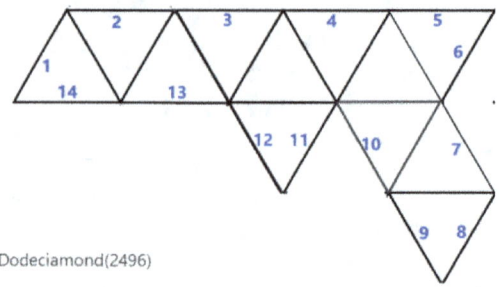

Dodeciamond (2496)

D12(3113)= (e1 e6)(e2 e3)(e4 e5)(e7 e14)(e8 e11) (e9 e10) (e12 e13)

12-Deltahedron (13/19)

Construct D12(2222) 1from Dodeciamond (2496).

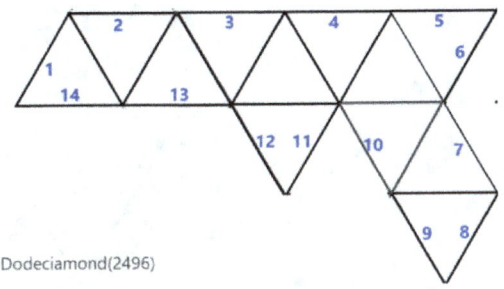

Dodeciamond (2496)

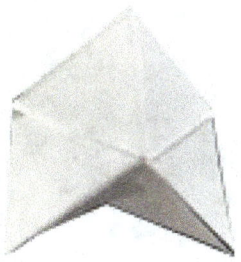

D12(2222)1= (e1 e8)(e2 e5)(e3 e4)(e6 e7)(e9 e12) (e10 e11) (e13 e14)

12-Deltahedron (14/19)

Construct D12(24002) from Dodeciamond (2496).

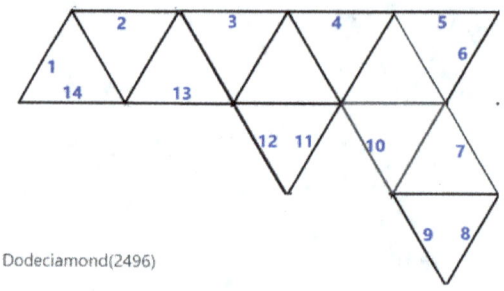

Dodeciamond (2496)

D12(24002)= (e1 e8)(e2 e7)(e3 e6)(e4 e5)(e9 e10) (e11 e14) (e12 e13)

12-Deltahedron (15/19)

Construct D12(1412) from Dodeciamond (2496).

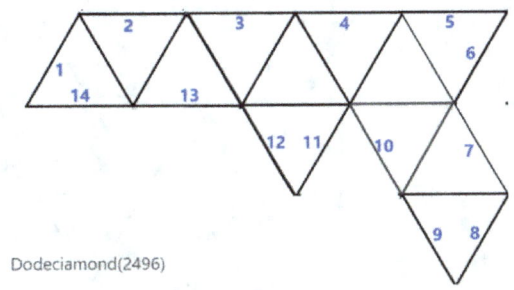

Dodeciamond (2496)

D12(1412)= (e1 e10)(e2 e9)(e3 e8)(e4 e5)(e6 e7) (e11 e14) (e12 e13)

12-Deltahedron (16/19)

Construct D12(31211) from Dodeciamond (2496).

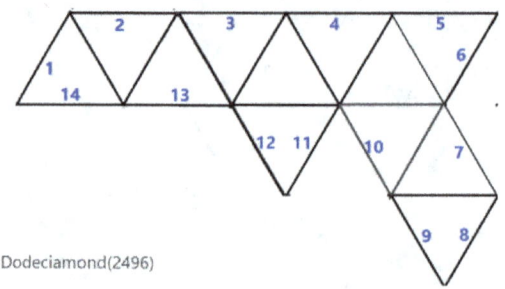

Dodeciamond (2496)

D12(31211) = (e1 e12)(e2 e9)(e3 e4)(e5 e8)(e6 e7) (e10 e11) (e13 e14)

12-Deltahedron (17/19)

Construct D12(2222)2 from Dodeciamond (2496).

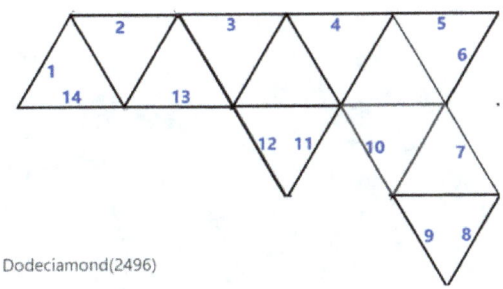

Dodeciamond (2496)

D12(2222)2= (e1 e12)(e2 e9)(e3 e8)(e4 e5)(e6 e7) (e10 e11) (e13 e14)

12-Deltahedron (18/19)

Thirteen (13) 12-Decahedra are

D12(0440), D12(1331), D12(1412), D12(2060), D12(2141)

D12(2222)1, D12(2222)2, D12(22301), D12(23111)

D12(24002), D12(3113), D12(31211), D12(4004)

12-Deltahedron (19/19)

We have found nine (9) 12-Deltahedron: D12(0440), D12(1331), D12(2222)2, D12(22301), D12(23111), D12(24002), D12(3113), D12(31211), D12(4004); 4 Non-deltahedral Graphs: D12(1412), D12(2060), D12(2141), D12(2222)1 .

12-Deitahedron has 12 faces (F), 18 edges (E), and 8 vertices (V).
All 13 12-Deltahedra satisfy Euler's Formula: F+V=E+2, F+V=12+8=20, E+2=18+2=20

CHAPTER 6

Apendices

A1 Tetriamonds

The number of Tetriamonds is 3, ie a(4) of sequence A000577 in the OEIS.

  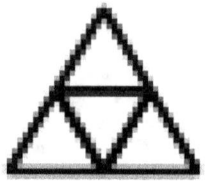

tetriamonds

A2 Hexiamonds

The number of Hexiamonds is 12, ie a(6) of sequence A000577 in the OEIS.

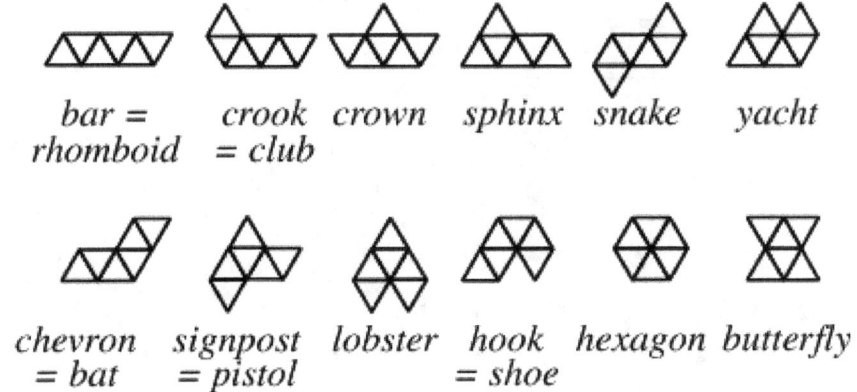

bar = rhomboid   crook = club   crown   sphinx   snake   yacht

chevron = bat   signpost = pistol   lobster   hook = shoe   hexagon   butterfly

A3 Octiamonds

The number of Octiamonds is 66, ie a(8) of sequence A000577 in the OEIS.

A4 Deciamonds

The number of Deciamonds is 448, ie a(10) of sequence A000577 in the OEIS.

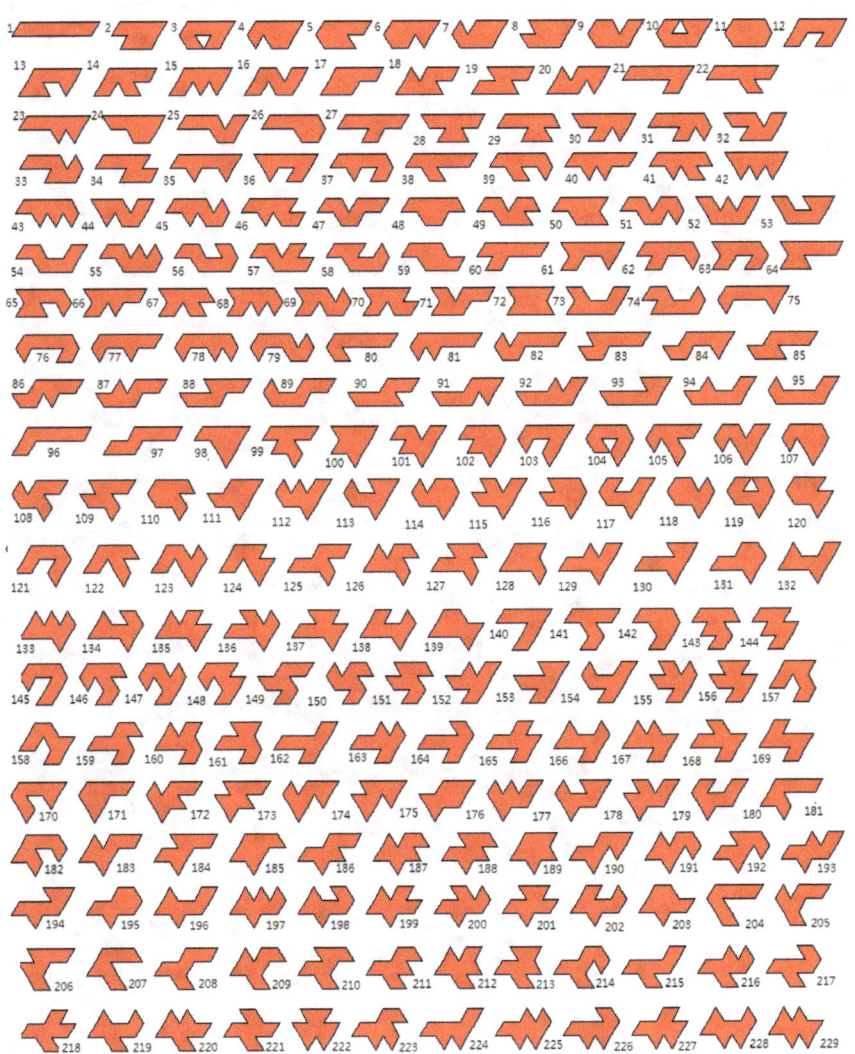

94

A5 Dodeciamonds (1/8)

The number of Dodeciamonds is 3334, ie a(12) of sequence A000577 in the OEIS.. The 3334 Dodeciamonds are divided into 8 Net Groups: 12e1, 12e2, 12e3, 12e4, 12e5, 12e6, 12e7, 12e8. Here is Net Group 12e1.

A5 Dodeciamonds (2/8)

12e2

A5 Dodeciamonds (3/8)

12e3

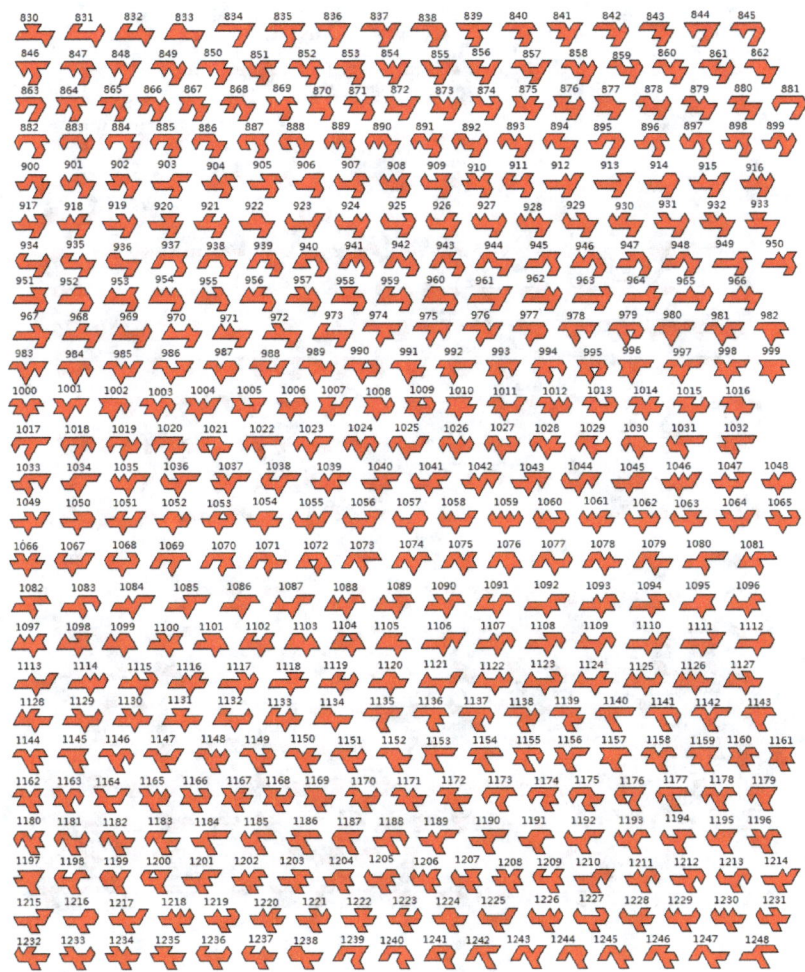

A5 Dodeciamonds (4/8)

12e4

A5 Dodeciamonds (5/8)

12e5

A5 Dodeciamonds (6/8)

12e6

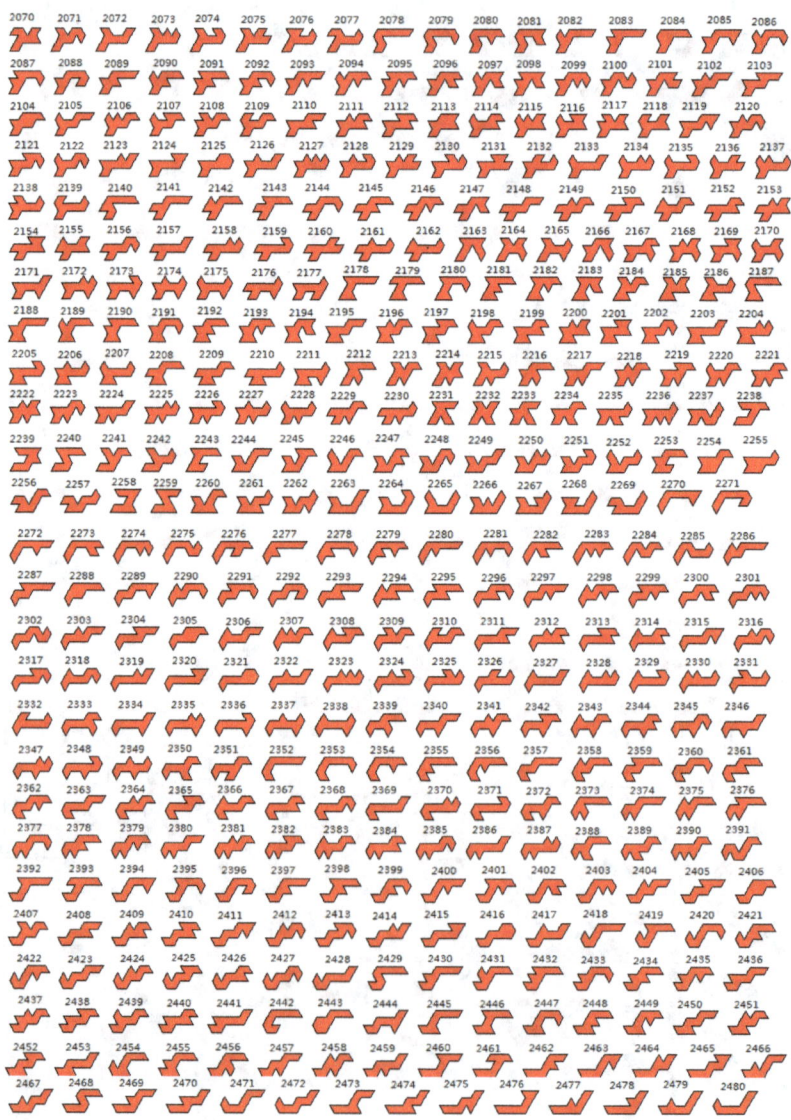

A5 Dodeciamonds (7/8)

12e7

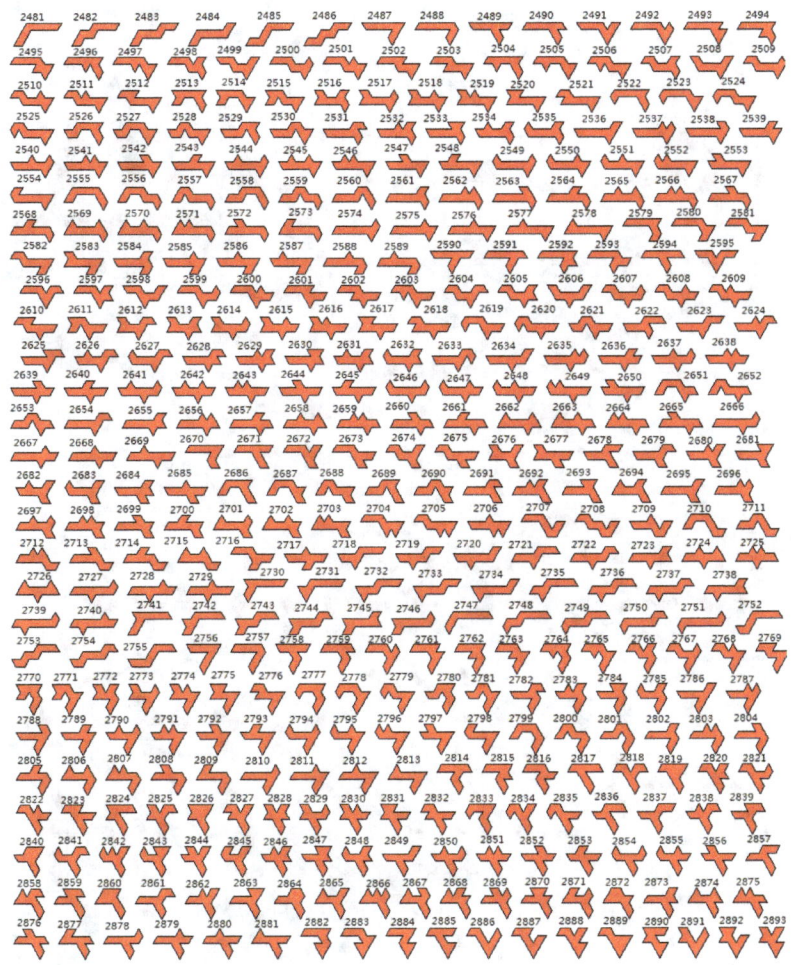

A5 Dodeciamonds (8/8)

12e8

www.ingramcontent.com/pod-product-compliance
Lightning Source LLC
Chambersburg PA
CBHW072103110526
44590CB00018B/3288